My FIRST Astrobiology Book

FOR KIDS

Sarah Wilson

First Edition April 2024

Content by Sarah Wilson

Book cover by Sarah Wilson

ISBN: 9798322862017

Published by Beeny Book Publishing

WELCOME TO THE CAPTIVATING WORLD OF ASTROBIOLOGY!

IN THIS BOOK, WE'LL EMBARK ON AN EXHILARATING JOURNEY THROUGH THE COSMOS, EXPLORING THE POTENTIAL FOR LIFE BEYOND OUR PLANET. FROM THE SEARCH FOR EXOPLANETS ORBITING DISTANT STARS TO THE STUDY OF EXTREMOPHILES THRIVING IN EARTH'S HARSHEST ENVIRONMENTS,

WE'LL DELVE INTO THE VAST AND INTRIGUING FIELD OF ASTROBIOLOGY. ALONG THE WAY, WE'LL MARVEL AT THE SHEER SIZE OF THE UNIVERSE, WITH ITS TRILLIONS OF GALAXIES AND COUNTLESS PLANETS, AND PONDER THE SIGNIFICANCE OF WATER AS A KEY INGREDIENT FOR LIFE AS WE KNOW IT.

JOIN US AS WE UNCOVER THE MYSTERIES OF THE COSMOS, CONTEMPLATE THE ORIGINS OF LIFE, AND SEEK TO UNDERSTAND OUR PLACE IN THE VAST EXPANSE OF THE UNIVERSE.

What is a Astrobiology?

ASTROBIOLOGY IS A FASCINATING FIELD OF SCIENCE THAT DELVES INTO THE PROFOUND QUESTION OF WHETHER LIFE EXISTS BEYOND EARTH. IT COMBINES VARIOUS SCIENTIFIC DISCIPLINES, INCLUDING ASTRONOMY, BIOLOGY, CHEMISTRY, AND PLANETARY SCIENCE, TO EXPLORE THE POTENTIAL FOR LIFE IN THE UNIVERSE. SCIENTISTS INVESTIGATE ENVIRONMENTS BOTH WITHIN OUR SOLAR SYSTEM AND BEYOND, SEARCHING FOR CONDITIONS THAT COULD SUPPORT LIFE AS WE KNOW IT.

Order of the Universe

THE UNIVERSE

A GALAXY

THE SOLAR SYSTEM

The Universe

THE UNIVERSE IS A VAST EXPANSE OF SPACE THAT CONTAINS BILLIONS OF GALAXIES. GALAXIES ARE DISTRIBUTED THROUGHOUT THE UNIVERSE AND STRETCH ACROSS UNIMAGINABLE DISTANCES. EVERY SPOT OF LIGHT IN THE IMAGE BELOW IS A GALAXY.

Galaxies

GALAXIES ARE VAST SYSTEMS OF STARS, GAS, AND DUST BOUND TOGETHER BY GRAVITY. THEY COME IN DIFFERENT SHAPES AND SIZES, FROM SMALL DWARFS TO GRAND SPIRALS AND ELLIPTICALS. EVERY GALAXY IS HOME TO BILLIONS OR EVEN TRILLIONS OF STARS.

IN THE BOUNDLESS REACHES OF THESE GALAXIES, STARS COME TOGETHER IN ARRANGEMENTS THAT RESEMBLE OUR OWN SOLAR SYSTEM, WITH PLANETS CIRCLING AROUND CENTRAL STARS.

Our Solar System

A SOLAR SYSTEM COMPRISES A SUN, A CENTRAL STAR, AND AN ARRAY OF CELESTIAL BODIES, INCLUDING PLANETS, MOONS, ASTEROIDS, AND COMETS, ALL BOUND TOGETHER BY GRAVITY IN ORBIT AROUND THE SUN. BELOW IS AN EXAMPLE OF OUR SOLAR SYSTEM.

NEPTUNE

SATURN

URANUS

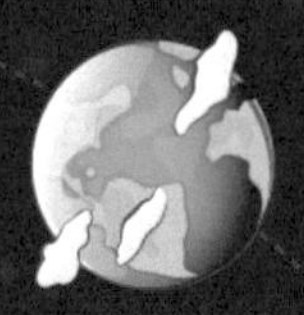

MARS

MERCURY

JUPITER

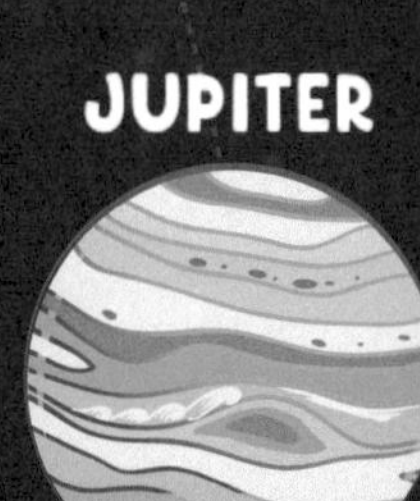

VENUS

BILLIONS TO TRILLIONS OF STARS. WITHIN THESE GALAXIES, STARS COME TOGETHER TO FORM SOLAR SYSTEMS, FEATURING A CENTRAL STAR ORBITED BY PLANETS, MOONS, ASTEROIDS, AND COMETS. WITH SO MANY GALAXIES AND THE POSSIBILITY OF MULTIPLE SOLAR SYSTEMS WITHIN EACH, THE TOTAL NUMBER OF PLANETS IN THE UNIVERSE IS IMMENSE, LIKELY NUMBERING IN THE TRILLIONS OR MORE. THIS ABUNDANCE OF GALAXIES AND SOLAR SYSTEMS SUGGESTS AN EQUALLY VAST ARRAY OF PLANETS AWAITING DISCOVERY, EACH POTENTIALLY HARBORING THE CONDITIONS NECESSARY FOR LIFE TO EMERGE.

The Earth

EARTH STANDS AS A TESTAMENT TO THE DELICATE BALANCE OF CONDITIONS THAT MAKE IT UNIQUELY HABITABLE IN THE VAST EXPANSE OF THE COSMOS. AMIDST THE INHOSPITABLE REALMS OF SPACE, EARTH EMERGES AS AN OASIS OF LIFE, OFFERING A RICH TAPESTRY OF ECOSYSTEMS TEEMING WITH BIODIVERSITY.

Water

WATER IS ESSENTIAL FOR LIFE AS WE KNOW IT BECAUSE IT SERVES AS A UNIVERSAL SOLVENT, ENABLING CRUCIAL CHEMICAL REACTIONS WITHIN LIVING ORGANISMS. IT PLAYS A VITAL ROLE IN VARIOUS BIOLOGICAL PROCESSES, SUCH AS METABOLISM AND THE TRANSPORT OF NUTRIENTS.

SINCE WATER IS ESSENTIAL FOR LIFE ON EARTH, ITS PRESENCE ON OTHER PLANETS OR MOONS COULD INDICATE THE POSSIBILITY OF HABITABLE CONDITIONS AND THE EXISTENCE OF LIFE FORMS SIMILAR TO THOSE FOUND ON OUR OWN PLANET.

The Habitable Zone

THE GOLDILOCKS ZONE, OFTEN REFERRED TO AS THE HABITABLE ZONE, IS A REGION AROUND A STAR WHERE CONDITIONS ARE NEITHER TOO HOT NOR TOO COLD, BUT JUST RIGHT FOR LIQUID WATER TO EXIST ON THE SURFACE OF A PLANET.

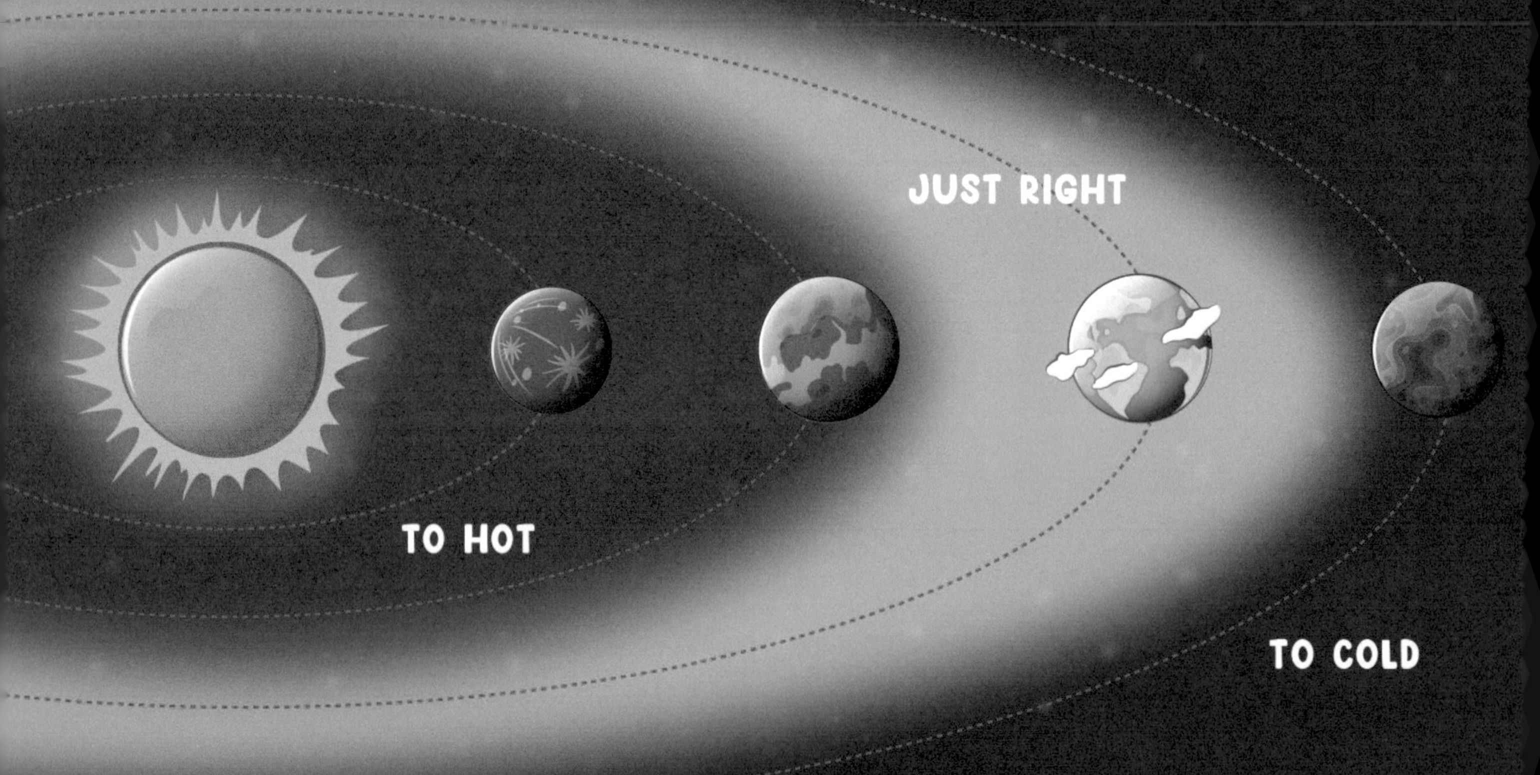

The Building Blocks of Life

ATOMS ARE TINY PARTICLES THAT MAKE UP EVERYTHING IN THE WORLD, INCLUDING LIVING THINGS. THEY COMBINE TO FORM MOLECULES, LIKE WATER, WHICH ARE ESSENTIAL FOR LIFE. UNDERSTANDING ATOMS HELPS US UNDERSTAND HOW LIVING ORGANISMS FUNCTION.

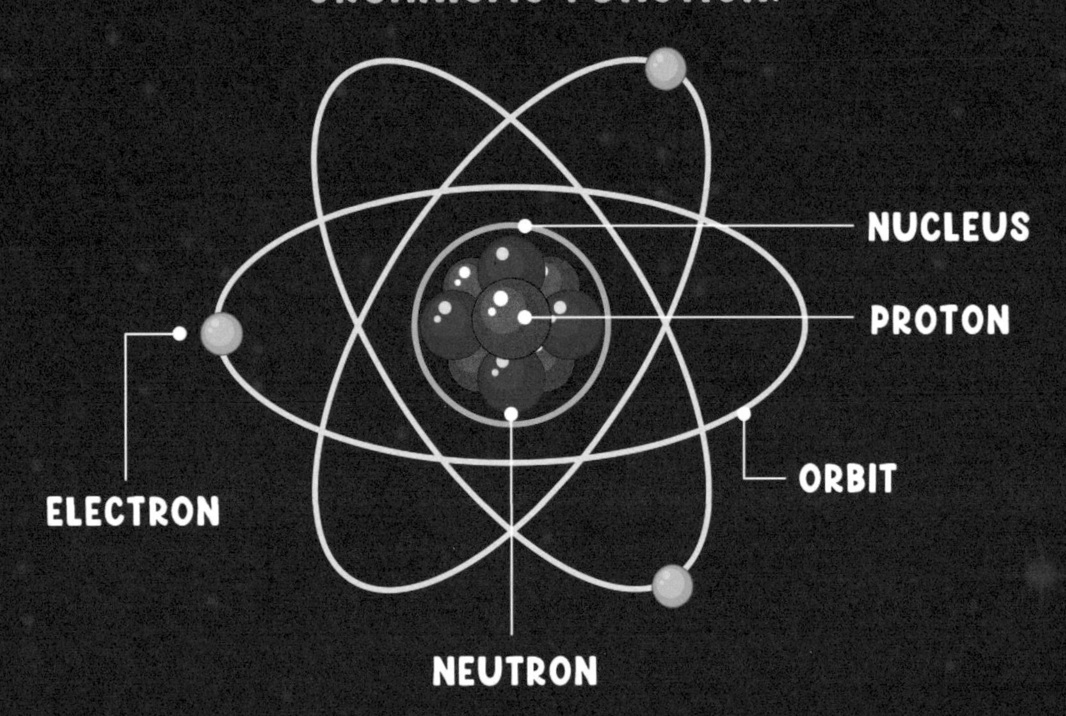

BY STUDYING THE ATOMS PRESENT ON OTHER PLANETS, SCIENTISTS CAN DETERMINE IF THE CONDITIONS NECESSARY FOR LIFE AS WE KNOW IT EXIST THERE.

Chemical Elements

CHEMICAL ELEMENTS ARE CRUCIAL IN THE SEARCH FOR LIFE ON OTHER PLANETS BECAUSE THEY FORM THE BASIS OF ALL MATTER, INCLUDING LIVING ORGANISMS. ELEMENTS LIKE CARBON, OXYGEN, NITROGEN, AND HYDROGEN ARE ESPECIALLY IMPORTANT AS THEY ARE ESSENTIAL FOR BUILDING ORGANIC MOLECULES NECESSARY FOR LIFE.

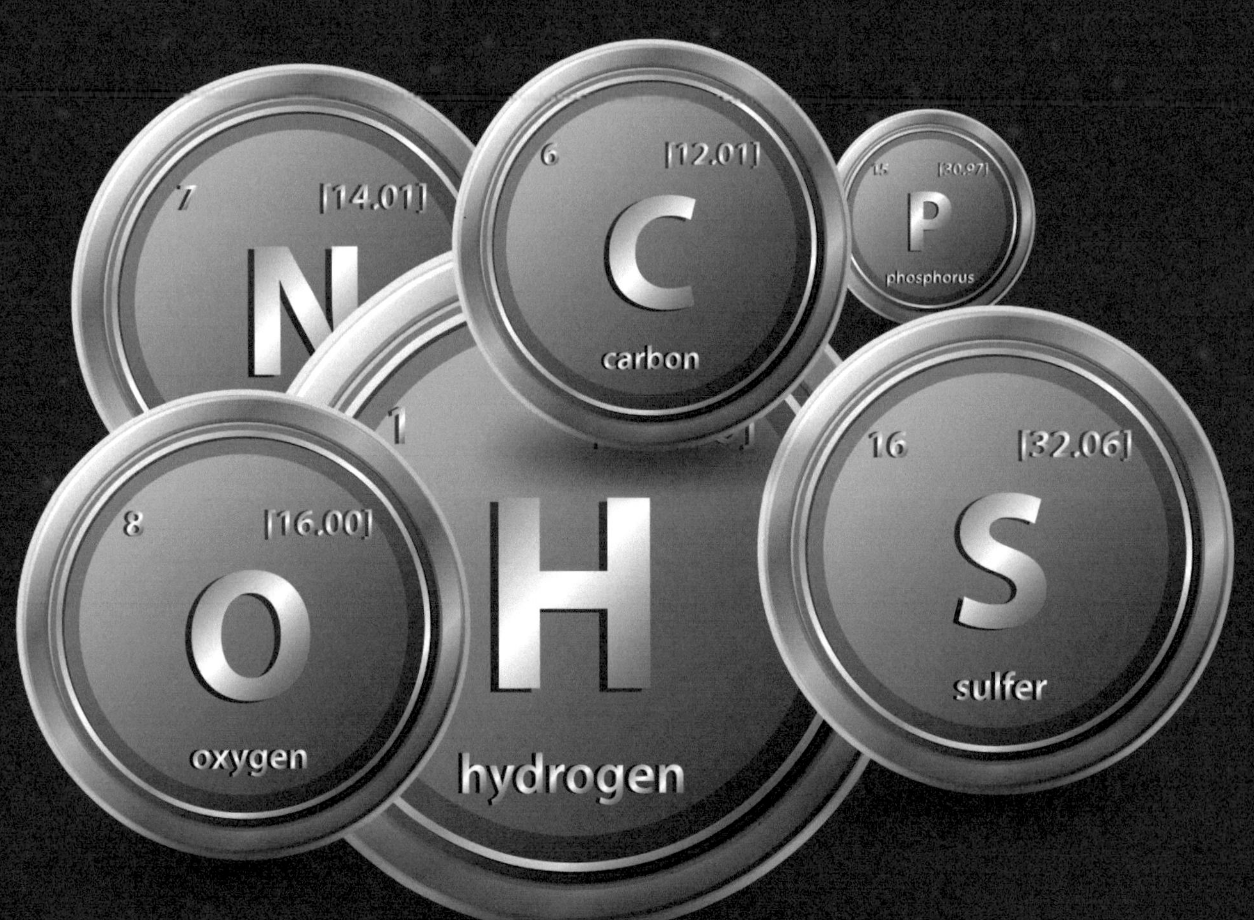

Photosynthesis

PHOTOSYNTHESIS IS VITAL IN THE SEARCH FOR LIFE ON OTHER PLANETS BECAUSE IT IS THE PROCESS BY WHICH MANY PLANTS AND SOME BACTERIA CONVERT SUNLIGHT INTO ENERGY, PRODUCING OXYGEN AS A BYPRODUCT. OXYGEN IS A KEY INDICATOR OF POTENTIALLY HABITABLE ENVIRONMENTS BECAUSE IT IS ESSENTIAL FOR THE RESPIRATION OF MANY ORGANISMS, INCLUDING HUMANS.

Atmosphere

THE ATMOSPHERE OF A PLANET IS IMPORTANT IN THE SEARCH FOR LIFE ELSEWHERE BECAUSE IT REGULATES TEMPERATURE, SHIELDS FROM HARMFUL RADIATION, AND CONTAINS GASES VITAL FOR LIFE, SUCH AS OXYGEN AND CARBON DIOXIDE. BY STUDYING THE COMPOSITION AND PROPERTIES OF A PLANET'S ATMOSPHERE, SCIENTISTS CAN ASSESS ITS POTENTIAL HABITABILITY AND LIKELIHOOD OF SUPPORTING LIFE.

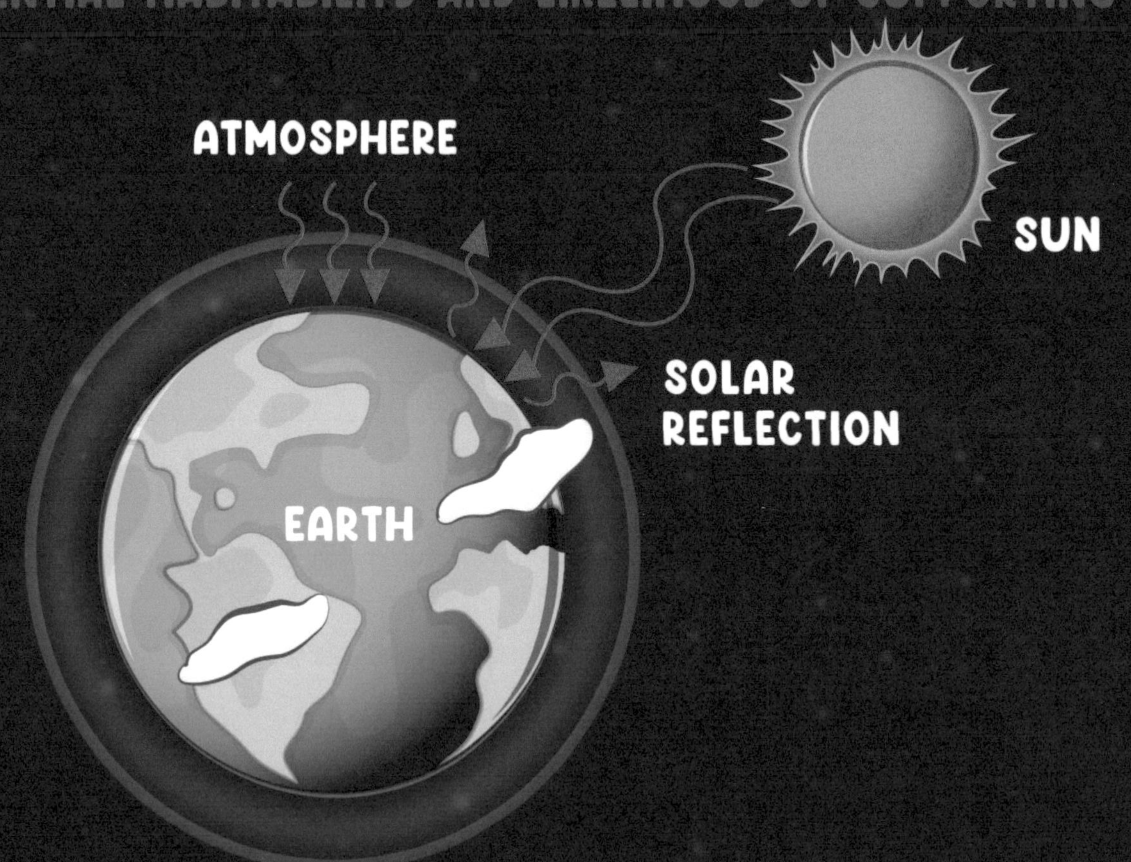

Structure of Earth

THE STRUCTURE OF OUR OWN PLANET, EARTH, PROVIDES VALUABLE INSIGHTS INTO UNDERSTANDING THE POTENTIAL HABITABILITY OF OTHER CELESTIAL BODIES. EARTH CONSISTS OF SEVERAL LAYERS, INCLUDING THE SOLID INNER CORE, THE LIQUID OUTER CORE, THE MANTLE, AND THE CRUST. EACH LAYER PLAYS A UNIQUE ROLE IN SHAPING EARTH'S SURFACE AND SUPPORTING LIFE.

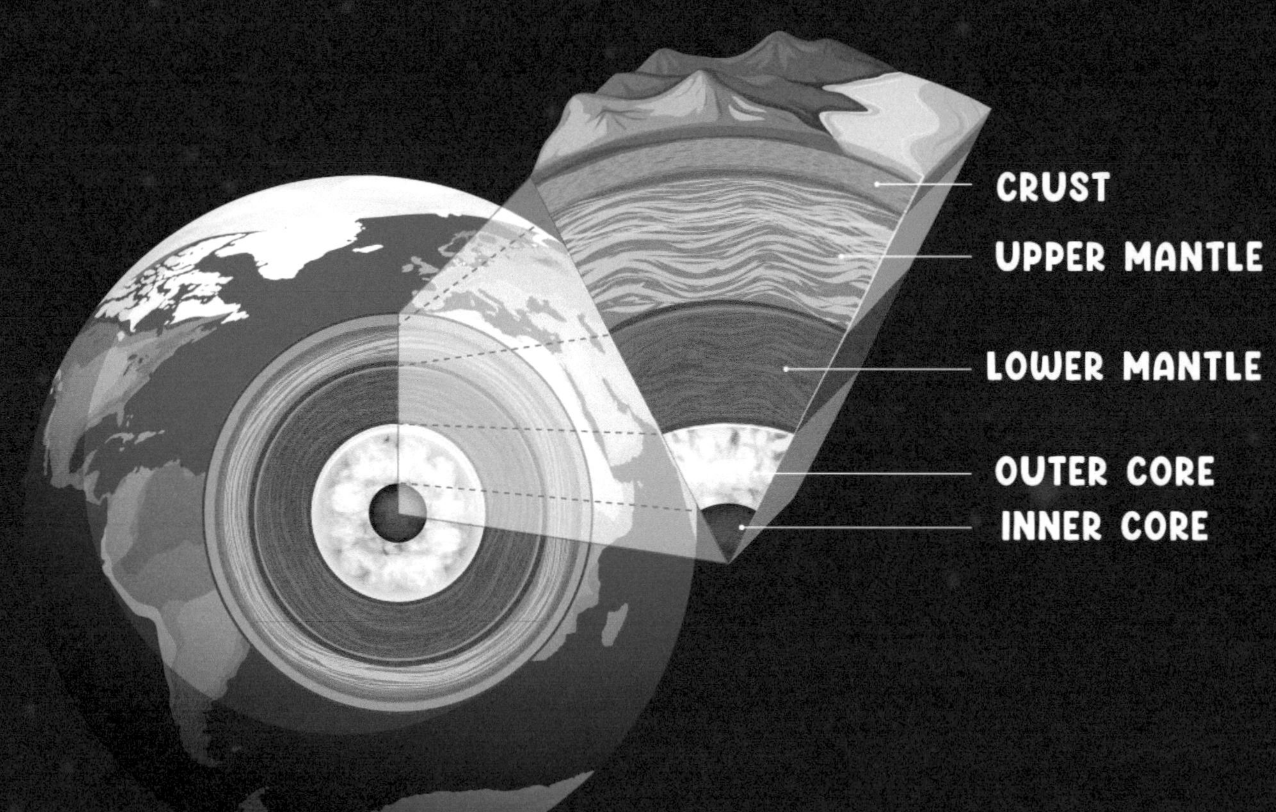

Extreme Enviroments

EXTREME ENVIRONMENTS ON EARTH, LIKE YELLOWSTONE AND DEEP-SEA HYDROTHERMAL VENTS, OFFER INSIGHTS INTO LIFE'S RESILIENCE IN HARSH CONDITIONS, PROVIDING VALUABLE CLUES FOR THE POTENTIAL EXISTENCE OF LIFE ELSEWHERE IN THE UNIVERSE.

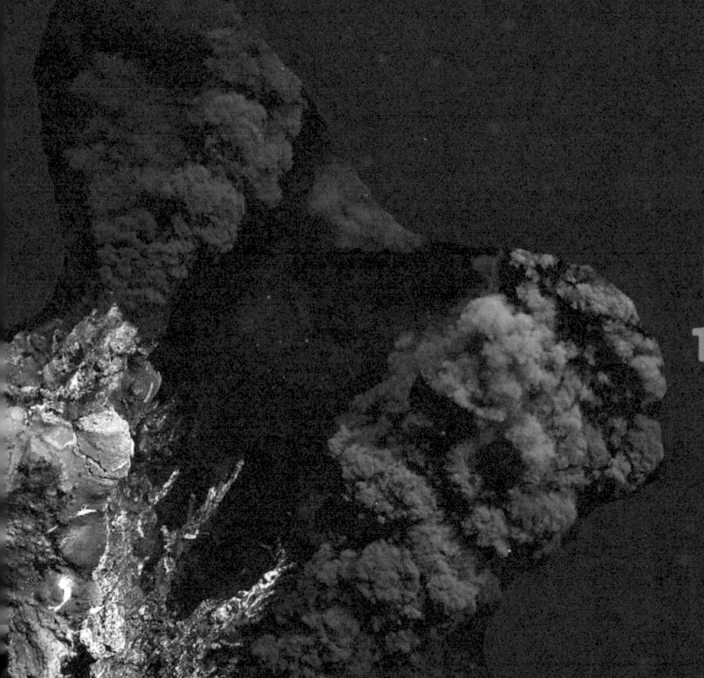

DESPITE FACING EXTREME TEMPERATURES, HIGHLY ACIDIC OR ALKALINE PH LEVELS, AND HIGH PRESSURE, MICROBIAL LIFE THRIVES IN THESE ENVIRONMENTS, DEMONSTRATING REMARKABLE ADAPTABILITY TO THE MOST CHALLENGING CONDITIONS. IMAGINABLE.

Extremophiles

EXTREMOPHILES ARE ORGANISMS THAT THRIVE IN EARTH'S MOST EXTREME ENVIRONMENTS, SUCH AS ACIDIC HOT SPRINGS, DEEP-SEA HYDROTHERMAL VENTS, AND POLAR ICE CAPS. THESE ENVIRONMENTS WERE ONCE THOUGHT TO BE INHOSPITABLE TO LIFE; HOWEVER, EXTREMOPHILES HAVE EVOLVED UNIQUE ADAPTATIONS THAT ALLOW THEM TO SURVIVE AND EVEN FLOURISH IN THESE HARSH CONDITIONS.

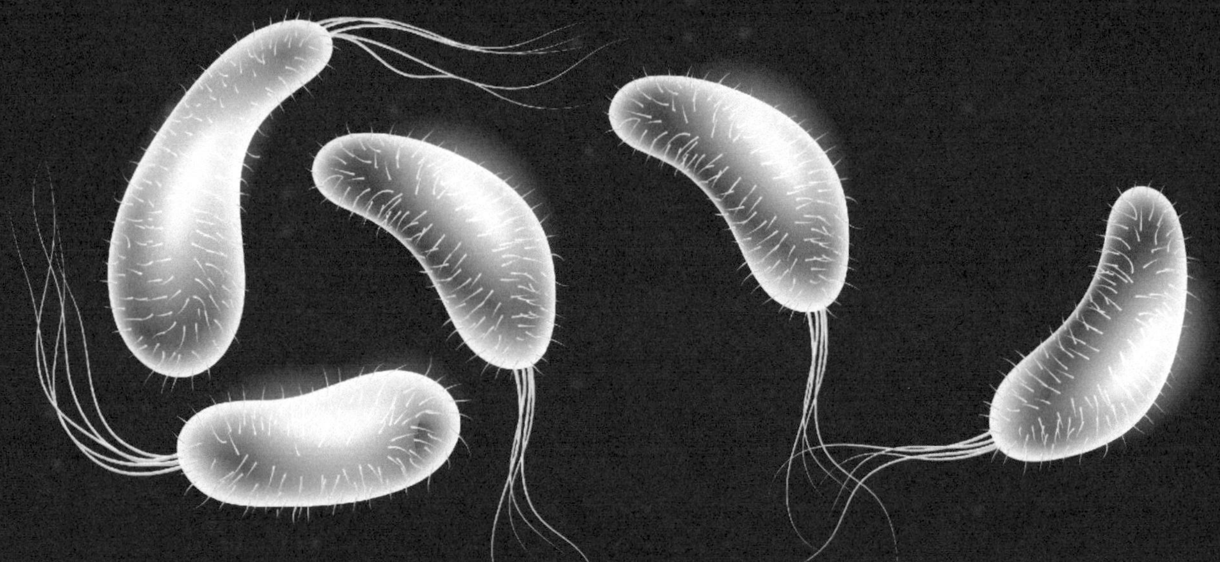

IN ASTROBIOLOGY, EXTREMOPHILES PLAY A CRUCIAL ROLE AS THEY PROVIDE INSIGHTS INTO THE POTENTIAL HABITABILITY OF EXTRATERRESTRIAL ENVIRONMENTS.

Europa

EUROPA IS ONE OF JUPITER'S MOONS WITH A SUBSURFACE OCEAN BENEATH ITS ICY SURFACE. ORGANIC COMPOUNDS HAVE BEEN FOUND ON EUROPA THROUGH OBSERVATIONS MADE BY SPACECRAFT MISSIONS LIKE GALILEO. THE DETECTION OF ORGANIC MOLECULES ON EUROPA SUGGESTS IT MIGHT HAVE THE INGREDIENTS FOR SUPPORTING MICROBIAL LIFE.

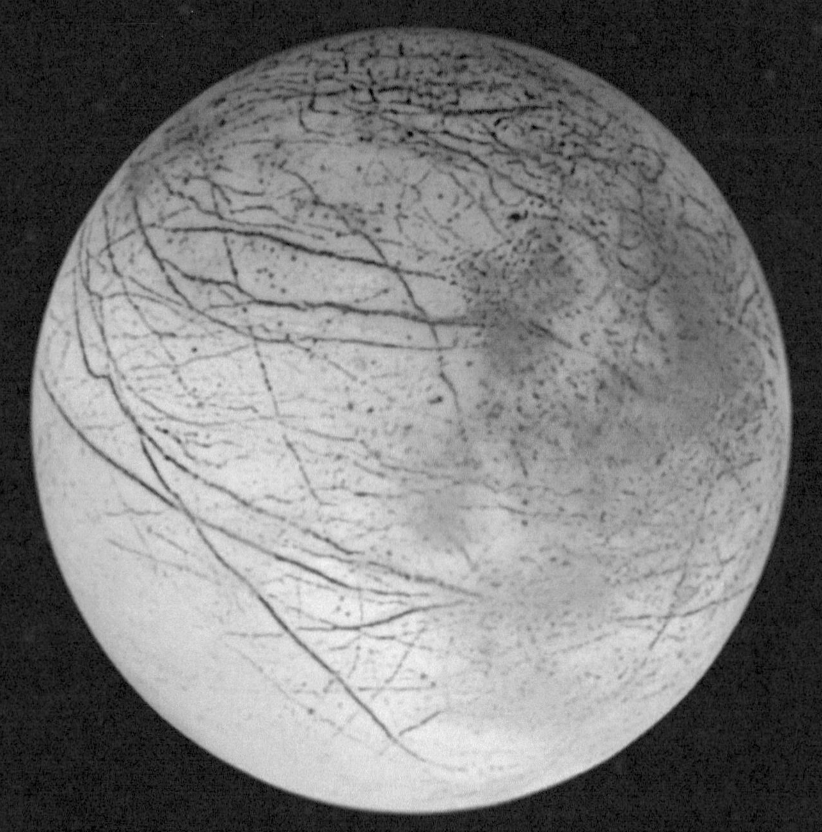

EUROPA IS ONLY ONE-FOURTH THE DIAMETER OF EARTH BUT ITS OCEAN MAY CONTAIN TWICE AS MUCH WATER AS EARTH'S.

Enceladus

ENCELADUS, A MOON OF SATURN, IS KNOWN FOR ITS SUBSURFACE OCEAN AND ERUPTING PLUMES OF WATER VAPOR. OBSERVATIONS FROM NASA'S CASSINI SPACECRAFT REVEAL A GLOBAL OCEAN BENEATH ITS ICY SURFACE, SUGGESTING POTENTIAL HABITABILITY. THE PLUMES HINT AT GEOLOGICAL ACTIVITY, POSSIBLY INCLUDING HYDROTHERMAL VENTS THAT COULD SUPPORT LIFE.

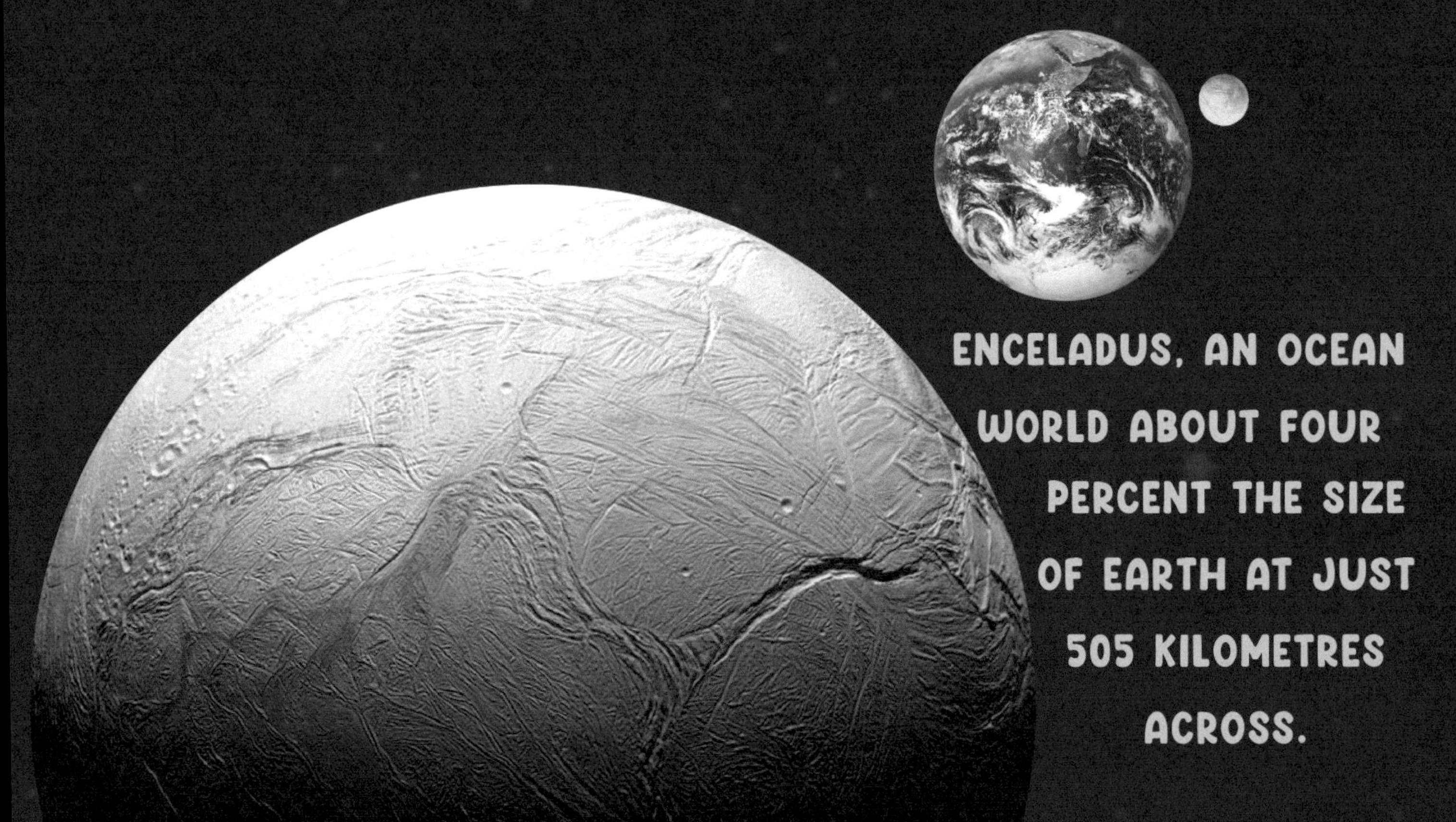

ENCELADUS, AN OCEAN WORLD ABOUT FOUR PERCENT THE SIZE OF EARTH AT JUST 505 KILOMETRES ACROSS.

Mars

MARS IS A KEY FOCUS DUE TO ITS POTENTIAL FOR PAST AND PRESENT LIFE. FEATURES LIKE ANCIENT RIVER VALLEYS AND EVIDENCE OF WATER SUGGEST MARS MAY HAVE BEEN HABITABLE IN THE PAST. EXPLORATION MISSIONS ALSO TARGET MARS' SUBSURFACE ENVIRONMENTS, WHERE MICROBIAL LIFE COULD POTENTIALLY SURVIVE TODAY.

Mars Rover

THE MARS ROVER IS A ROBOTIC VEHICLE DESIGNED TO EXPLORE THE SURFACE OF MARS. EQUIPPED WITH CAMERAS, SPECTROMETERS, AND DRILLS, THESE ROVERS STUDY MARTIAN TERRAIN, ATMOSPHERE, AND GEOLOGY. THEY HAVE ADVANCED OUR UNDERSTANDING OF THE RED PLANET BY SEARCHING FOR SIGNS OF PAST MICROBIAL LIFE, CHARACTERIZING MARS' ENVIRONMENT, AND PAVING THE WAY FOR FUTURE HUMAN EXPLORATION.

How do you find life?

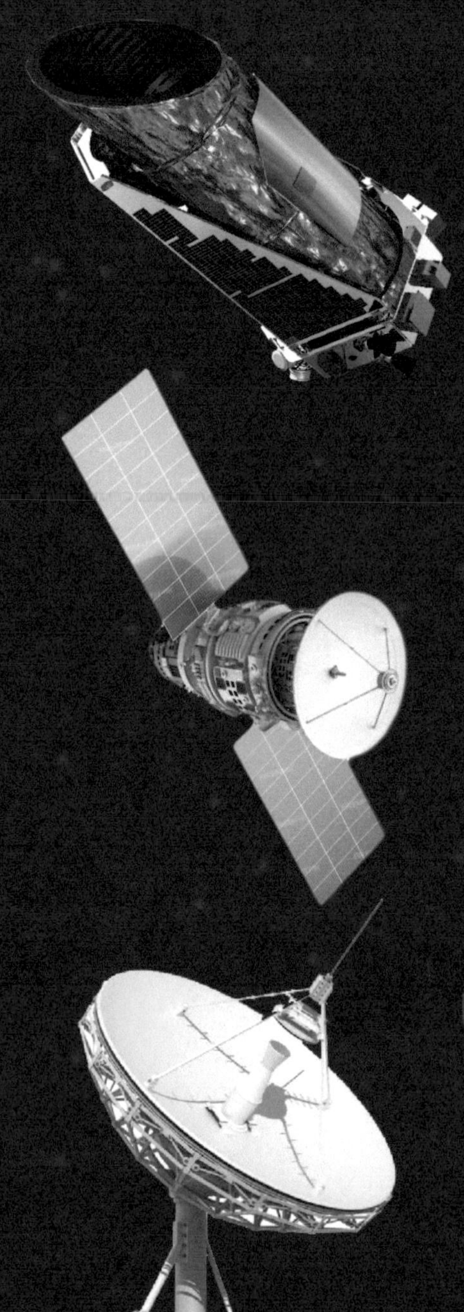

LAUNCHED IN 2009, NASA'S KEPLER SPACE TELESCOPE IDENTIFIED THOUSANDS OF EXOPLANETS, REVOLUTIONIZING ASTROBIOLOGY AND GUIDING THE SEARCH FOR HABITABLE WORLDS BEYOND OUR SOLAR SYSTEM.

SATELLITES ENABLE SCIENTISTS TO STUDY CELESTIAL BODIES, SUCH AS PLANETS, MOONS, AND ASTEROIDS, IN DETAIL, HELPING TO IDENTIFY POTENTIAL HABITATS FOR LIFE.

RADIO TELESCOPES SCAN THE COSMOS FOR RADIO SIGNALS THAT MAY INDICATE THE PRESENCE OF EXTRATERRESTRIAL LIFE. THESE INSTRUMENTS SEARCH FOR POTENTIAL SIGNALS ORIGINATING FROM DISTANT PLANETS OR GALAXIES.

Exoplanets

EXOPLANETS ARE PLANETS ORBITING STARS BEYOND OUR SOLAR SYSTEM. THOUSANDS HAVE BEEN DISCOVERED, REVEALING A DIVERSE ARRAY OF WORLDS. THESE FINDINGS OFFER INSIGHTS INTO THE PREVALENCE OF PLANETS IN THE UNIVERSE AND THE POTENTIAL FOR HABITABLE ENVIRONMENTS BEYOND EARTH.

SIMILAR IN SIZE

EARTH

KEPLER-1649C

KEPLER-1649C IS AN EXOPLANET. IT ORBITS A RED DWARF STAR IN THE HABITABLE ZONE, WHERE CONDITIONS MIGHT SUPPORT LIQUID WATER. THIS MAKES IT A PROMISING CANDIDATE FOR HOSTING LIFE BEYOND OUR SOLAR SYSTEM.

Quiz

WHAT IS THE GOLDILOCKS ZONE?

A. A REGION IN SPACE WHERE ASTEROIDS ARE MOST COMMON

B. A REGION AROUND A STAR WITH CONDITIONS SUITABLE FOR LIFE

C. THE AREA AROUND A BLACK HOLE WHERE LIGHT CANNOT ESCAPE

D. A REGION IN THE MILKY WAY WITH THE MOST STARS

WHAT IS THE NAME OF THIS CELESTIAL BODY?

A. MARS

B. EUROPA

C. JUPITER

D. ENCELADUS

WHAT IS AN EXTREMOPHILE?

A. A TYPE OF EXOPLANET WITH EXTREME WEATHER PATTERNS

B. AN ORGANISM THAT THRIVES IN EXTREME ENVIRONMENTS

C. AN ASTEROID FOUND IN THE OUTER SOLAR SYSTEM

D. A MOON ORBITING A GAS GIANT PLANET

Quiz

WHICH OF THE FOLLOWING COSMIC STRUCTURES IS THE LARGEST?

A. THE UNIVERSE B. A GALAXY C. A SOLAR SYSTEM?

ENCELADUS IS A MOON ORBITING WHICH PLANET?

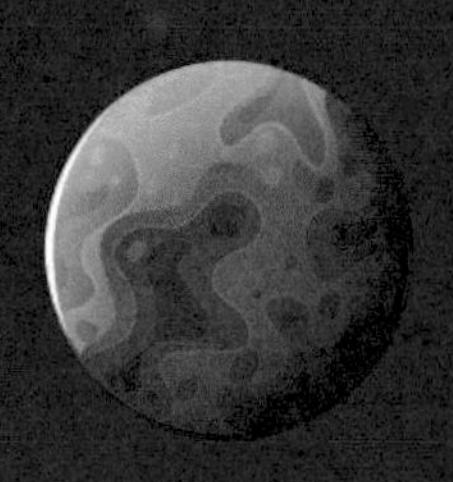

A. JUPITER B. SATURN C. EARTH D. MARS

WHAT IS THE KEPLER SPACE TELESCOPE KNOWN FOR?

A. STUDYING BLACK HOLES

B. OBSERVING GAMMA-RAY BURSTS

C. IDENTIFYING EXOPLANETS

D. MONITORING ASTEROIDS

Answers

WHAT IS THE GOLDILOCKS ZONE?

~~A. A REGION IN SPACE WHERE ASTEROIDS ARE MOST COMMON~~

B. A REGION AROUND A STAR WITH CONDITIONS SUITABLE FOR LIFE

~~C. THE AREA AROUND A BLACK HOLE WHERE LIGHT CANNOT ESCAPE~~

~~D. A REGION IN THE MILKY WAY WITH THE MOST STARS~~

WHAT IS THE NAME OF THIS CELESTIAL BODY?

~~A. MARS~~

B. EUROPA

~~C. JUPITER~~

~~D. ENCELADUS~~

WHAT IS AN EXTREMOPHILE?

~~A. A TYPE OF EXOPLANET WITH EXTREME WEATHER PATTERNS~~

B. AN ORGANISM THAT THRIVES IN EXTREME ENVIRONMENTS

~~C. AN ASTEROID FOUND IN THE OUTER SOLAR SYSTEM~~

~~D. A MOON ORBITING A GAS GIANT PLANET~~

Answers

WHICH OF THE FOLLOWING COSMIC STRUCTURES IS THE LARGEST?

A. THE UNIVERSE ~~B. A GALAXY~~ ~~C. A SOLAR SYSTEM?~~

ENCELADUS IS A MOON ORBITING WHICH PLANET?

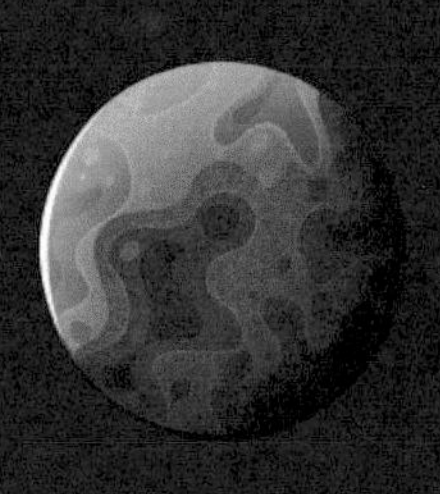

~~A. JUPITER~~ B. SATURN ~~C. EARTH~~ ~~D. MARS~~

WHAT IS THE KEPLER SPACE TELESCOPE KNOWN FOR?

~~A. STUDYING BLACK HOLES~~

~~B. OBSERVING GAMMA-RAY BURSTS~~

C. IDENTIFYING EXOPLANETS

~~D. MONITORING ASTEROIDS~~

AS WE CONCLUDE OUR JOURNEY THROUGH ASTROBIOLOGY, WE'RE LEFT WITH A PROFOUND APPRECIATION FOR THE POTENTIAL OF LIFE BEYOND OUR PLANET. ASTROBIOLOGY, THE STUDY OF LIFE IN THE UNIVERSE, ENCOMPASSES A WIDE RANGE OF DISCIPLINES, FROM THE EXPLORATION OF EXOPLANETS TO THE SEARCH FOR EXTREMOPHILES ON EARTH. THIS INTERDISCIPLINARY FIELD NOT ONLY SHEDS LIGHT ON THE ORIGINS AND EVOLUTION OF LIFE BUT ALSO DEEPENS OUR UNDERSTANDING OF THE COSMOS AND OUR PLACE WITHIN IT.

YOUR INSIGHTS AND REFLECTIONS ARE INVALUABLE TO US AS WE CONTINUE TO EXPLORE THE MYSTERIES OF ASTROBIOLOGY. WE INVITE YOU TO SHARE YOUR THOUGHTS AND CONSIDER LEAVING A REVIEW. THANK YOU FOR JOINING US ON THIS EXTRAORDINARY EXPLORATION OF THE COSMOS!

Made in United States
Troutdale, OR
12/01/2024